Nanotechnology Made Simple

The Future of Science, Demystified

By

Peter Kattan

Book Bound Press

https://web.facebook.com/BookboundPress/

Preface

In the vast universe of scientific exploration, few areas are as transformative, intriguing, and promising as nanotechnology. The word itself— a blend of "nano," meaning billionth, and "technology," signifying applied science— conjures images of a future where the impossible becomes routine. Yet for many, nanotechnology remains a mystifying domain, cloaked in technical jargon and seemingly reserved for elite researchers and scientists. **Nanotechnology Made Simple: The Future of Science, Demystified** is a response to that mystique, crafted to bring this groundbreaking field to the curious mind in a clear, accessible, and engaging way.

This book emerges from a deep conviction: understanding nanotechnology is no longer optional in the 21st century. Its impact permeates countless facets of our lives, from the medications that heal us to the electronics that connect us, the energy solutions that power our homes, and the sustainable practices that ensure our planet's future. The world is increasingly shaped by this "small" science that operates at the molecular and atomic levels, with "big" implications for humanity.

The journey through this book is designed to empower you. Whether you're a student exploring career options, a professional curious about the latest scientific advances, or simply an enthusiast seeking to grasp the essence of one of the most dynamic fields of our era, this guide has something for you. Each chapter builds on the last, painting a vivid picture of what nanotechnology is, how it works, and why it matters.

We begin by laying the foundation with an introduction to the history, scope, and importance of nanotechnology. From there, we delve into the intricate science that governs the nanoscale, demystifying the unique properties of materials and the ingenious techniques used to manipulate them. The book then explores the vast applications of nanotechnology, from revolutionary medical treatments and environmental breakthroughs to transformative changes in electronics, energy, and even agriculture.

Safety, ethics, and societal implications also occupy a central place in our narrative. As with any powerful tool, nanotechnology demands responsible stewardship. This book does not shy away from addressing the challenges it brings— regulatory hurdles, potential risks, and ethical dilemmas—ensuring readers have a balanced understanding of both its promises and perils.

What sets this book apart is its future-facing perspective. We illuminate not only the current applications of nanotechnology but also the emerging trends and innovations poised to shape the decades ahead. From artificial intelligence accelerating nanomaterial design to international collaborations solving global challenges, the horizon of possibilities is as inspiring as it is expansive.

Finally, **Nanotechnology Made Simple** concludes with a call to action. It invites readers to see themselves as participants in the nanotechnology revolution, emphasizing that the breakthroughs of tomorrow depend on the curiosity, creativity, and contributions of individuals today. Whether as researchers, policymakers, educators, or informed citizens, we all have a role to play in steering nanotechnology toward a brighter, more equitable future.

This book is not just about nanotechnology; it is about demystifying the science that defines our future and inspiring a sense of wonder and responsibility in its readers. May this journey leave you not only informed but also empowered, ready to explore the boundless opportunities of a world shaped by the smallest of scales.

Welcome to the nanoscale. Let the journey begin.

Peter I. Kattan   December 2024

Introduction

In a world buzzing with tech breakthroughs, nanotechnology shines bright as a game-changer. Picture this: manipulating materials at the atomic level, reshaping the very essence of matter to tackle some of humanity's biggest issues. This isn't some far-off dream; it's happening right now, and it's closer than you think.

Since scientists first peered into the nanoscale, the potential has been mind-blowing. This field's evolved fast, flipping how we look at medicine, electronics, environmental care, and even farming. With each new discovery, nanotech isn't just changing science; it's shaking up our everyday lives in ways we're just starting to grasp.

Think about tiny particles designed to deliver lifesaving meds straight to cancer cells, cutting down on side effects while ramping up effectiveness. Imagine holding a device in your hand that packs the power of a supercomputer— all thanks to the miniaturization magic of nanotech. Picture a future where clean water is standard, not a luxury, as cutting-edge filtration systems use nanomaterials to purify our resources.

But hey, with great power comes some serious responsibility. As we dive deeper into nanotechnology, we've gotta face the ethical dilemmas and risks that come with it. How do we make sure these innovations are safe for folks and the planet? What rules do we need to set up to guide the smart use of nanomaterials? These questions are just as important as the scientific breakthroughs, and they'll steer the future of this field for years to come.

In this look at nanotech, we'll trace its history, highlighting the key moments that led to today's advancements. We'll break down the science behind the nanoscale, showing how materials act differently at this tiny level and how we can tap into these traits for real-world uses. From healthcare to energy, electronics to agriculture, the upcoming chapters will shed light on the countless ways nanotechnology is set to change our lives.

As we gaze into the future, we'll also explore how nanotech intersects with artificial intelligence. This combo promises to speed up research and deepen our understanding of materials at the atomic level. Together, they're ready to unlock new possibilities that once felt like sci-fi.

This isn't just a science book; it's a call to jump into a revolution that's already in motion. Whether you're a seasoned pro, a curious student, or just someone fascinated by tech's potential, you'll find insights and inspiration here. You'll see how you can be part of this thrilling ride, helping shape a future where innovation and responsibility go hand in hand.

So, come along as we dive into the fascinating world of nanotechnology, where the tiny becomes huge, and the future is only limited by our imagination. Let's unlock the secrets of the nanoscale together and dream up a world transformed by the possibilities ahead. The adventure starts now.

Table of Contents

Chapter 1

Introduction to Nanotechnology

Let's embark on an exhilarating journey into the realm of nanotechnology! Don't let the term intimidate you; it's not as complicated as it sounds. At its essence, nanotechnology is about manipulating matter at an incredibly small scale—specifically, at the nanoscale, which is one billionth of a meter. To put that into perspective, think about the size of a single grain of sand compared to an entire beach. It's a mind-boggling concept, isn't it? When we begin to tinker with materials at this minuscule level, we open the door to a world of endless possibilities. We're dealing with atoms and molecules, the fundamental building blocks of everything that surrounds us.

So, what's the scope of this captivating field? It's vast and varied, touching upon numerous sectors, from medicine to electronics, energy to materials science. Picture nanotechnology as the Swiss Army knife of science—versatile and brimming with potential. Imagine tiny particles engineered to deliver medications precisely to the source of an illness, enhancing treatment effectiveness while minimizing side effects. Or think about how nanomaterials can revolutionize our gadgets, making them lighter, stronger, and more efficient. The possibilities are limitless, and as we peel back the layers, we find ourselves standing at the very forefront of innovation.

Let's take a moment to stroll down memory lane and examine some historical milestones that paved the way for the nanotechnology we know today. This story begins long before the term "nanotechnology" was even coined. In the 1950s, physicist Richard Feynman delivered a groundbreaking lecture titled "There's Plenty of Room at the Bottom." He envisioned a future where we could manipulate individual atoms— an idea that, at the time, seemed like pure science fiction. Yet, that spark ignited a fire, sowing the seeds of what would eventually become nanotechnology.

Fast forward to the 1980s, when scientists began developing techniques to visualize and manipulate materials at the nanoscale. The invention of the scanning tunneling microscope (STM) was a game-changer, allowing researchers to see and move individual atoms. It was akin to giving a child a brand-new toy; the excitement was palpable! With each advancement, the field blossomed, and in 2000, the National Nanotechnology Initiative was launched in the United States, signaling that this wasn't just a fleeting trend—it was a full-blown revolution.

Now, why does all this matter? Why should we care about nanotechnology? The answer is simple: it's reshaping our world. In modern science, nanotechnology serves as a cornerstone of innovation. It's not merely a buzzword; it's a catalyst for change. From enhancing renewable energy sources to developing more effective cancer treatments, the impact of nanotechnology is profound. It holds the promise of addressing some of the most pressing challenges we face today—climate change, healthcare disparities, and even the quest for sustainable resources.

As we navigate this thrilling terrain, it's crucial to remember that the beauty of nanotechnology lies in its potential to transform lives. With the right applications, we can significantly enhance the quality of life for millions of people around the globe. Whether through advanced medical therapies that target diseases at their root or materials that help reduce our carbon footprint, nanotechnology stands as a beacon of hope in a world that often feels overwhelmed by problems.

As we embark on this journey through the world of nanotechnology, keep your mind open and your curiosity ignited. This isn't just a technical field; it's a vibrant landscape filled with opportunities for growth and discovery. The future is bright, and you're about to explore a realm that's not just changing science—it's changing lives. Embrace the excitement, and let's dive in!

Understanding the Nanoscale

To truly appreciate the marvels of nanotechnology, we must first grasp the concept of the nanoscale. Imagine a world where you can see and manipulate individual atoms, where the tiniest changes can lead to monumental advancements. At the nanoscale, materials behave differently than they do at larger scales, and this opens up a treasure trove of possibilities.

Think of it this way: if you were to scale up a nanometer to the size of a basketball, then a meter would be equivalent to the distance from the basketball court to the moon! This perspective highlights just how small the nanoscale is. At this level, materials can exhibit unique properties, such as increased strength, lighter weight, and enhanced chemical reactivity. These characteristics are what make nanotechnology so exciting and impactful.

Let's consider an example to illustrate this point. Gold, a material we often think of as heavy and dense, behaves quite differently at the nanoscale. When gold particles are reduced to the nanoscale, they can appear red or purple instead of their traditional metallic hue. This change in color is due to the way light interacts with the particles at this size. Such phenomena are not just fascinating; they can be harnessed for various applications, including medical imaging and targeted drug delivery.

In essence, the nanoscale allows us to manipulate materials in ways that were previously unimaginable. By understanding and harnessing these unique properties, we can create innovative solutions that address real-world challenges. Whether it's developing new materials for construction, improving the efficiency of solar cells, or creating targeted therapies for diseases, the potential is vast.

Applications of Nanotechnology

Now that we have a grasp on the nanoscale, let's explore some of the exciting applications of nanotechnology. The impact of this field can be felt across numerous industries, and the possibilities are only beginning to be realized.

1. Medicine: One of the most promising applications of nanotechnology lies in the field of medicine. Imagine tiny nanoparticles designed to deliver drugs directly to cancer cells, minimizing damage to healthy tissue. This targeted approach can lead to more effective treatments with fewer side effects. Additionally, nanotechnology can be utilized in diagnostics, enabling earlier detection of diseases through advanced imaging techniques.

2. Electronics: The electronics industry is undergoing a transformation thanks to

nanotechnology. With the ability to create smaller, more efficient components, we can develop faster and more powerful devices. From smartphones to computers, nanotechnology is paving the way for the next generation of technology. Think about how nanomaterials can enhance battery life, making our devices last longer and perform better.

3. Energy: As we face the challenges of climate change, nanotechnology offers innovative solutions for energy production and storage. Nanomaterials can improve the efficiency of solar cells, making renewable energy sources more viable. Additionally, advancements in nanotechnology can lead to the development of better batteries, allowing for more efficient energy storage and usage.

4. Environmental Remediation: Nanotechnology has the potential to play a significant role in addressing environmental

issues. Nanoparticles can be used to clean up pollutants in soil and water, breaking down harmful substances and restoring ecosystems. This application not only benefits the environment but also promotes public health and safety.

5. Materials Science: The field of materials science is being revolutionized by nanotechnology. By manipulating materials at the nanoscale, we can create stronger, lighter, and more durable materials. This has applications in various industries, including construction, aerospace, and automotive. Imagine building structures that are not only more resilient but also more energy-efficient!

6. Food and Agriculture: Nanotechnology is also making strides in the food and agriculture sectors. From improving food packaging to enhancing crop yields, the applications are diverse. Nanomaterials can help create

packaging that extends the shelf life of food products, reducing waste and promoting sustainability.

As you can see, the applications of nanotechnology are far-reaching and transformative. Each of these areas holds the potential to improve our quality of life and address some of the most pressing challenges we face today. The key takeaway here is that nanotechnology is not just a scientific curiosity; it's a powerful tool that can drive innovation and positive change.

Challenges and Ethical Considerations

While the potential of nanotechnology is exciting, it's essential to approach this field with a sense of responsibility. As with any emerging technology, there are challenges and ethical considerations that must be addressed.

1. Safety: One of the primary concerns surrounding nanotechnology is safety. As we manipulate materials at the nanoscale, we must ensure that they do not pose risks to human health or the environment. Rigorous testing and regulation are necessary to assess the safety of nanomaterials before they are widely used.

2. Environmental Impact: While nanotechnology has the potential to benefit the environment, it can also have unintended consequences. For example, the release of nanoparticles into ecosystems could disrupt natural processes. It's crucial to study and understand the long-term effects of nanotechnology on the environment to mitigate any negative impacts.

3. Ethical Implications: As we delve deeper into the capabilities of nanotechnology, ethical questions arise. For instance, how do we ensure

equitable access to these advancements? Will certain populations benefit more than others? Addressing these questions is vital to ensure that the benefits of nanotechnology are shared fairly.

4. Public Perception: The public's understanding of nanotechnology is still evolving. Misinformation and misconceptions can lead to fear and resistance to new technologies. It's essential to engage in open dialogue with the public, providing accurate information and addressing concerns to foster trust and acceptance.

By acknowledging and addressing these challenges, we can harness the power of nanotechnology responsibly. It's not just about what we can do; it's about what we should do. As we move forward, let's commit to ethical practices and prioritize the well-being of individuals and the planet.

The Future of Nanotechnology

As we look to the future, the possibilities for nanotechnology are boundless. With ongoing research and innovation, we can expect to see even more groundbreaking applications emerge. Imagine a world where diseases are detected and treated at their earliest stages, where renewable energy sources power our homes and cities, and where materials are designed to be both strong and sustainable.

The future of nanotechnology is not just about technological advancements; it's about improving lives. It's about creating solutions that enhance our quality of life and address global challenges. As you embark on your own journey in the world of nanotechnology, keep this vision in mind. You have the power to contribute to this exciting field and make a difference in the lives of others.

Visualize your success as a writer in this space. Picture your work inspiring others to explore the wonders of nanotechnology, igniting curiosity and passion. You have a unique voice and perspective that can resonate with readers, guiding them through the complexities of this field. Embrace your role as a storyteller, sharing not just facts and figures, but also the human experiences that underscore the importance of nanotechnology.

As you write, remember that your journey is just as important as the destination. Celebrate each step along the way, whether it's completing a chapter, conducting research, or connecting with fellow enthusiasts. Each small win brings you closer to your goal of crafting a compelling narrative that captivates and informs.

In conclusion, the world of nanotechnology is an exciting frontier filled with potential. It's a field that invites curiosity, innovation, and exploration. As you delve into this subject, keep your heart open and your mind engaged. Embrace the challenges and celebrate the triumphs. The future is bright, and you're about to embark on a journey that not only changes science but also transforms lives. Let's get started!

Chapter 2

The Science Behind Nanotechnology

Alright, let's break down the science of nanotechnology into something a bit more digestible. You ready? Here we go!

When we chat about nanotechnology, we're diving into this wild, teeny-tiny world that's way more significant than it seems. Think of it like looking through a microscope and stumbling upon a whole new universe. At the center of this adventure is the nanoscale, which is measured in nanometers—yeah, that's one billionth of a meter. Mind-blowing, right? But don't sweat it; we're gonna take it slow, like a dance at prom, one step at a time.

Understanding the nanoscale is kinda like learning the rules of a new game. You've gotta know the field before you can play. This is where the real magic happens. Materials start acting all kinds of different at this scale. Picture a basketball player shrinking down to the size of a marble. Everything changes—how they dribble, shoot, score—you name it! That's what materials do at the nanoscale; they show off some pretty unique properties that can lead to breakthroughs in tech, medicine, and beyond.

So, why should we care? Well, the significance of the nanoscale is huge. At this level, materials can become stronger, lighter, and more reactive. Imagine a steel beam that's ten times tougher but half the weight. That's a game changer! This shift in properties unlocks a treasure chest of innovation possibilities. We're talking everything from sturdier buildings to super-efficient solar panels. It's like opening a toolbox filled with gadgets you didn't even know existed!

Let's dive deeper into what makes materials tick at the nanoscale. When you shrink materials down, their surface area blows up compared to their volume. This means more atoms are hanging out in the open, which cranks up their reactivity. Think of it like a sponge soaking up water—the more surface area, the more water it can soak up. In nanotech, this can be a game changer for stuff like drug delivery systems, where tiny particles are designed to hit specific cells in the body. It's like having a delivery truck that knows exactly where to drop off its goodies!

Now, here's where it gets even cooler: quantum effects. When materials hit the nanoscale, they start showing off some quantum properties that lead to surprising behaviors. For example, gold nanoparticles can turn red or purple instead of that classic shiny yellow. It's like a magic trick that defies what we expect! This funky behavior can be used in all sorts of applications, from sensors to imaging technologies.

Let's talk about how we actually manipulate these tiny materials. This is where the artistry kicks in. Scientists and engineers have whipped up some pretty slick techniques to shape and mold materials at this minuscule level. One of the big ones is lithography, where light creates patterns on a material. It's kinda like using a stencil to paint a design—only we're working with stuff a million times smaller than the tip of a pencil!

Then there's self-assembly. Picture a bunch of kids on a playground forming a circle without anyone telling them what to do. They just know where to go! This happens because of the interactions between particles, driven by forces like van der Waals and hydrogen bonding. Tapping into self-assembly can lead to complex nanostructures for drug delivery, electronics, and even energy storage.

Another cool method is chemical vapor deposition. This lets us create thin films of materials by depositing them from a gas phase onto a surface. Think of it like frosting a cake— layer by layer, you build something beautiful and functional. This technique is a big deal in making semiconductors and nanostructured materials, paving the way for advancements in electronics and photonics.

As we dig deeper into nanotech, it's crucial to remember this journey isn't just about science. It's about how these tiny innovations can flip our lives upside down. Each breakthrough at the nanoscale has the potential to change how we live, work, and interact with the world. So, keep your eyes peeled! The future is bright, and it's brimming with opportunities just waiting for us to grab 'em.

Now, let's not forget about the ethical side of things. With great power comes great responsibility, right? As we push the boundaries of what's possible with nanotechnology, we've gotta think about the implications. Are we ready for the changes this tech can bring? Will it help everyone, or just a select few? These are questions we need to tackle as we venture into this brave new world.

And speaking of implications, let's chat about some real-world applications of nanotechnology. It's not just a bunch of scientists in lab coats mixing potions. Nah, this stuff is already changing lives! Take medicine, for example. Nanoparticles are being used in targeted drug delivery, which means meds can go straight to the problem area without affecting the rest of the body. It's like sending in a sniper instead of a bomb squad!

In the realm of electronics, nanotech is making waves too. We're seeing faster processors, smaller devices, and better batteries. Ever heard of quantum dots? These little guys are revolutionizing displays, making colors pop like never before. It's like going from black-and-white TV to full-on IMAX in your pocket!

And let's not forget about energy. Nanotechnology is playing a massive role in creating more efficient solar panels and batteries. Imagine a solar panel that can absorb sunlight even on cloudy days. That's not science fiction; that's nanotech in action!

But hold up—let's not get too carried away with the hype. There are still challenges to face. We need to figure out how to scale up production without losing quality. Plus, we've gotta ensure that these materials are safe for humans and the environment. Nobody wants to be a guinea pig for the next big thing, right?

Now, as we wrap this up, let's circle back to the essence of nanotechnology. It's all about unlocking the potential of materials at this tiny scale. The unique properties that emerge can lead to groundbreaking applications across various fields. By using innovative techniques to manipulate these materials, we're crafting solutions that can enhance our quality of life.

So, as you set out on your own writing journey, keep in mind that the science behind nanotechnology isn't just a bunch of facts—it's a story filled with exploration, creativity, and the promise of a better tomorrow. Embrace the challenge, keep pushing forward, and let your curiosity lead the way. You've got this!

And hey, don't forget to have a little fun along the way. Science can be serious business, but it's also a wild ride. So, whether you're a student, a curious mind, or just someone who stumbled upon this topic, remember that every great discovery starts with a question. So, what's yours?

Chapter 3

Nanomaterials: Types and Applications

Alright, let's dive into the fascinating world of nanomaterials. Imagine cracking open a treasure chest filled with tiny wonders—smaller than a human hair, but packing a punch in terms of potential. Seriously, these little guys are game-changers. So, let's break it down and see what's what.

First up, we've got three main categories of nanomaterials: nanoparticles, nanocomposites, and nanotubes. Each one's got its own unique flair, kinda like superheroes in a comic book.

Nanoparticles are the little soldiers in the nanotechnology brigade. They can be crafted from metals, polymers, or ceramics, and they come in all sorts of shapes—think tiny spheres, cubes, or rods. Their small size gives them a massive surface area compared to their volume, which amps up their reactivity. You'll find them everywhere, from your skincare products to sunscreens, boosting performance like nobody's business. They're like the secret ingredient in a killer recipe—hard to notice, but totally essential.

Next on the list are nanocomposites. These bad boys are the superheroes of the material world. They blend two or more different materials at the nanoscale to create something new and stronger. Picture mixing chocolate and peanut butter. Individually, they're great, but together? A flavor explosion! These nanocomposites are used in all sorts of stuff—lighter yet tougher packaging, construction materials that can weather the storm, you name it.

And then there are nanotubes. These cylindrical structures made from carbon atoms are like the muscle of the nanomaterial world— super strong and lightweight. Imagine a material that's six times stronger than steel but way lighter. Crazy, right? They're already shaking things up in the electronics field, popping up in transistors and sensors. Plus, researchers are looking into their potential for energy storage solutions. Talk about versatility!

Now, let's switch gears and chat about how these nanomaterials are turning the electronics and energy storage worlds upside down. In electronics, they're totally revolutionizing our devices. Thanks to nanomaterials, we can miniaturize tech, cramming more power into smaller packages. Think about your smartphone—what if I told you the magic inside it comes from nanomaterials? Yup! They boost conductivity and efficiency, making our gadgets faster and more reliable.

When it comes to energy storage, we're on the brink of some serious breakthroughs. Nanomaterials can supercharge batteries and supercapacitors, allowing for lightning-fast charging and longer-lasting power. Imagine charging your phone in just a few minutes and it lasting for days. That's the dream, folks! Researchers are diving into nanostructured electrodes and electrolytes to make this a reality, pushing the envelope on what we thought was possible.

Now, let's not overlook the role of nanomaterials in medicine and healthcare. This is where it gets really exciting. Nanomaterials are paving the way for groundbreaking innovations in drug delivery systems. Picture a tiny capsule, smaller than a grain of salt, delivering medication right to the cells that need it most. That's the magic of targeted drug delivery. It cuts down on side effects and amps up effectiveness—total win-win!

But wait, there's more. Nanomaterials are also jazzing up imaging techniques, giving doctors a clearer view inside our bodies. With nanoparticles in play, we can enhance contrast in imaging scans, making it easier to catch diseases early. It's like giving doctors a super-powered pair of glasses that help them spot issues before they get outta hand.

And let's not forget the potential for targeted cancer therapies. Researchers are exploring ways to use nanomaterials to deliver cancer-fighting drugs straight to tumor cells, sparing the healthy ones. It's kinda like a guided missile, hitting the target without collateral damage. This kind of precision could change the game for cancer treatment, making it more effective and less invasive.

So, what's the bottom line? The world of nanomaterials is bursting with potential, from electronics to healthcare. Each type brings its own unique benefits and applications, and together, they're driving innovation like never before.

As you embark on your own journey into nanotechnology, remember this: every small step can lead to monumental changes. You're part of a larger story—one that's unfolding right before our eyes. Keep that curiosity alive, and who knows what you might stumble upon next?

In the grand scheme of things, nanomaterials aren't just about their size; they're about their impact. They're transforming industries, improving lives, and shaping the future. And you, my friend, are on the brink of this exciting revolution. So, buckle up and get ready to explore the wonders of nanotechnology!

Now, let's dive deeper into each category and explore some specific examples and applications that showcase the magic of nanomaterials.

Starting with nanoparticles, they come in various forms, each with its own superpowers. Gold nanoparticles, for instance, are all the rage in medical diagnostics. Their unique optical properties allow them to be used in imaging and sensing applications. Imagine a tiny gold particle that can help doctors see what's happening in the body at a molecular level. It's like having a backstage pass to the inner workings of our cells!

Then we've got silver nanoparticles, which are known for their antimicrobial properties. They're popping up in everything from wound dressings to food packaging. The idea is that these little guys can help keep things clean and safe by killing bacteria. It's like having a tiny bodyguard that protects you from germs.

And don't forget about quantum dots. These tiny semiconductor particles are super cool because they emit light of specific colors when exposed to UV light. They're being used in displays and lighting technologies, making screens brighter and more vibrant. It's like adding a splash of color to your favorite painting—totally transforms the whole picture!

Moving on to nanocomposites, let's talk about how they're making a splash in the construction industry. You know those concrete structures that seem to last forever? Well, adding nanomaterials to concrete can enhance its strength and durability. It's like giving concrete a workout—making it tougher and more resilient against the elements.

And in the world of packaging, nanocomposites are being used to create materials that are not only lightweight but also incredibly strong. This means we can reduce waste while ensuring our products are protected. It's a win for the environment and a win for consumers.

Now, onto nanotubes. These cylindrical wonders are already making waves in the electronics world. They're being used in transistors, which are the building blocks of all electronic devices. Thanks to their exceptional electrical conductivity, nanotubes can help create faster and more efficient devices. Imagine your computer running at lightning speed—thanks to these tiny tubes!

But it doesn't stop there. Nanotubes are also being explored for their potential in energy storage. Researchers are looking into using them in batteries to improve performance and longevity. It's like upgrading your phone battery to last twice as long without needing a charge—who wouldn't want that?

Now, let's take a moment to reflect on the impact of nanomaterials in the healthcare sector. The possibilities are endless! For instance, scientists are working on using nanoparticles to improve the effectiveness of vaccines. By delivering the vaccine directly to the immune cells, we could enhance the body's response, making vaccines even more powerful. It's like giving your immune system a turbo boost!

And how about using nanomaterials for tissue engineering? Researchers are exploring ways to create scaffolds using nanomaterials that can help regenerate damaged tissues.

Imagine being able to heal injuries faster or even grow new organs. It's like science fiction becoming reality!

Let's not forget about the environmental benefits of nanomaterials. They can play a crucial role in water purification, helping to remove contaminants and pollutants from our drinking water. Picture a tiny filter that can trap harmful substances, ensuring we have clean and safe water to drink. It's a simple yet powerful application that can have a massive impact on public health.

As we wrap this up, it's clear that nanomaterials are not just a passing trend—they're shaping the future in ways we're just beginning to understand. From revolutionizing electronics to transforming healthcare, the potential is limitless.

So, what's the takeaway? Embrace the curiosity. Dive into the world of nanotechnology and explore the possibilities. You never know what you might discover. After all, in the grand tapestry of innovation, every thread counts. And who knows? You might just be the one to unravel the next big breakthrough in nanomaterials. So, keep your eyes peeled and your mind open—the future's looking bright!

Chapter 4

Nanotechnology in Medicine

When you think about medicine, what pops into your head? White coats, stethoscopes, maybe the chaos of a hospital? Sure, that's part of it. But there's a whole microscopic world out there that's just waiting to revolutionize the way we heal—nanotechnology. This little marvel is shaking things up in medicine like a snow globe on a frosty morning.

Let's kick things off with nanoparticles. Now, I know you're probably wondering, "What the heck is a nanoparticle?" Picture this: tiny delivery trucks zipping through your bloodstream, transporting medicine straight to where it's needed. Traditional drug delivery? It's like trying to hit a bullseye from a mile away. But with nanoparticles, we're talkin' precision. These little champs can be designed to release their cargo only when they reach their target cells. That means fewer side effects and a whole lot more effectiveness. It's like having a secret weapon in your medicine cabinet!

Think about a cancer patient. With standard treatments, they often get hit hard with side effects because those drugs don't discriminate—they just go for everything in sight. But nanoparticles? They're the smart cookies in the bunch. They can be engineered to recognize cancer cells, latching on like a magnet and delivering the drug right where it needs to go. This targeted approach not only amps up the treatment's effectiveness but also spares healthy cells from unnecessary damage. It's a total game changer, folks.

Now, let's switch gears and chat about imaging techniques boosted by nanotech. Ever heard the phrase, "seeing is believing"? In the medical world, that's gospel. Imaging is crucial for diagnosis and treatment planning. But here's the kicker: traditional imaging methods can sometimes miss the finer details. That's where nanotechnology struts in like a superhero with a magnifying glass.

Nanoparticles can serve as contrast agents in imaging techniques like MRI and CT scans. They jazz up the clarity and detail of the images, allowing doctors to see what's really going on inside the body. Think of it as going from a black-and-white TV to high-def. Suddenly, those fuzzy images become crystal clear, revealing tumors and abnormalities that might've slipped through the cracks. This leap forward means earlier detection, better treatment planning, and, ultimately, improved patient outcomes.

But hold on, there's even more! Let's dig into the potential for targeted cancer therapies. We've already touched on how nanoparticles can deliver drugs straight to cancer cells, but that's just scratching the surface. Researchers are diving deep into using nanotech to not just deliver drugs but also to supercharge the body's immune response against cancer. It's like giving your immune system a turbo boost!

Imagine a world where cancer treatments are personalized, tailored to each person's unique biology. That's the dream, and it's getting closer to reality thanks to nanotechnology. By combining nanoparticles with immunotherapy, scientists are figuring out how to make the body recognize and attack cancer cells more effectively. This means fewer side effects, less suffering, and a better quality of life for patients. It's a win-win, if you ask me!

Now, let's take a moment to visualize the future. Picture a hospital where every patient gets a customized treatment plan that uses nanotechnology. Doctors use enhanced imaging to pinpoint exactly where the issue lies, then deploy nanoparticles to deliver targeted therapies that minimize side effects. Patients walk out of the hospital feeling hopeful, knowing they're getting the best care possible.

But it's not just about the tech itself; it's about the stories behind it. Each advancement in nanotechnology carries the promise of transformation for real people. You probably know someone who's been affected by cancer or another serious illness. The potential for targeted therapies could mean the difference between life and death for them. That's some powerful stuff!

Now, I get it—you might be thinkin', "This all sounds amazing, but what's the catch?" And you're right to ask that. With great power comes great responsibility. As we embrace the wonders of nanotechnology in medicine, we gotta tread carefully. We need to consider the ethical implications, the safety of these nanoparticles, and the long-term effects on patients. It's a delicate balance, but one we must navigate with care and foresight.

So, what can you do with all this info? First off, keep your curiosity alive. Dive into the research, read up on the latest advancements, and stay informed. The more you know, the better equipped you'll be to engage in conversations about the future of medicine. Second, if you're in a position to advocate for change—whether that's in healthcare policy, research funding, or community education—take that step. Your voice matters!

Lastly, remember that you're part of this revolution, whether you realize it or not. Every time you share what you've learned, you're helping to spread awareness about the incredible potential of nanotechnology in medicine. You might inspire someone else to explore this field or even pursue a career in healthcare or research. The ripple effect can be profound!

As we wrap up this exploration of nanotechnology in medicine, I want you to hold onto this thought: the future is bright. With every tiny advancement, we're paving the way for a healthier tomorrow. So, keep dreaming big, stay curious, and remember that you have the power to make a difference. The journey of discovery is just beginning, and who knows what incredible breakthroughs lie ahead? Together, let's embrace the possibilities and watch as nanotechnology transforms lives, one nanoparticle at a time.

Let's get a bit more into the nitty-gritty of nanoparticles. These aren't just random bits of matter; they're specifically designed structures that can vary in size, shape, and surface properties. Imagine a Swiss Army knife—each tool serves a different purpose, right? Well, nanoparticles can be tailored to carry drugs, detect diseases, or even deliver genetic material. They're versatile little buggers!

Now, the science behind this is pretty wild. Nanoparticles can be made from various materials, including metals, lipids, and polymers. Each material has its perks. For instance, gold nanoparticles are super popular in medical applications because they're biocompatible and can be easily modified. They can also absorb light and convert it to heat, which can be used to destroy cancer cells. How cool is that?

Let's talk about drug delivery a bit more. In the past, when doctors prescribed medication, it was like tossing a bunch of darts at a board and hoping they hit the target. But with nanoparticles, it's like having a laser-guided missile. They can be engineered to respond to specific triggers, like pH changes or temperature shifts, ensuring that the drug is released only when it reaches the right spot. This precision not only boosts the drug's effectiveness but also minimizes the chances of those nasty side effects we all dread.

And don't forget about the role of nanotechnology in diagnostics. Early detection is key in treating diseases like cancer. Traditional diagnostic tools might miss subtle signs, but with the help of nanoparticles, doctors can catch things way earlier. Imagine a tiny nanoparticle that can bind to a specific biomarker associated with a disease. When it does, it can emit a signal that doctors can pick up on. It's like having a tiny detective in your bloodstream, constantly on the lookout for trouble.

Now, let's get into some real-life examples. Take cancer treatment, for instance. Researchers have been working on using nanoparticles to deliver chemotherapy drugs directly to tumors. This means that instead of the drugs circulating throughout the entire body and wreaking havoc on healthy cells, they're going straight to the source. One study showed that using nanoparticles to deliver a specific drug resulted in a 90% reduction in tumor size compared to traditional methods. That's a jaw-dropper!

And it's not just cancer. Nanotechnology is making waves in treating autoimmune diseases, cardiovascular issues, and even infections. Imagine a future where we can target and treat diseases with pinpoint accuracy. That's the kind of world I want to live in!

But let's not get too carried away without addressing the elephant in the room: safety. Just because something is small doesn't mean it's harmless. We need to be cautious about how these nanoparticles interact with our bodies.

There's still a lot we don't know about the long-term effects of introducing these tiny particles into our systems. It's crucial for researchers to conduct thorough studies to ensure that these advancements don't come with hidden dangers.

So, what does the future hold? It's hard to say for sure, but the possibilities are endless. With ongoing research, we're bound to see more breakthroughs that could change the face of medicine as we know it. Picture this: a world where doctors can create personalized treatment plans using a patient's unique genetic makeup, all thanks to nanotechnology. That's not just science fiction; it's becoming a reality!

As we move forward, let's also think about accessibility. It's one thing to have cutting-edge technology; it's another to make sure everyone can benefit from it. We need to advocate for policies that ensure equitable access to these treatments, especially for underserved communities. After all, the goal is to improve health for everyone, not just a select few.

So, what can you do? Stay informed. Talk to your friends and family about these advancements. The more we share knowledge, the more we can push for positive change in the healthcare system. And if you're passionate about this field, consider getting involved. Whether it's pursuing a career in medicine, research, or policy advocacy, your efforts can help shape the future of healthcare.

In closing, let's remember that we're all part of this journey. The advancements in nanotechnology are just the beginning. With every tiny step, we're moving closer to a future where medicine is more effective, personalized, and accessible. So, keep dreaming big, stay curious, and embrace the incredible potential of nanotechnology in medicine. Together, we can make a difference—one nanoparticle at a time!

Chapter 5

Environmental Applications of

Nanotechnology

Imagine a world where clean water flows freely, pollution is a thing of the past, and our materials are sustainable and efficient. Sounds like a dream, right? Well, let me tell you, that dream is inching closer to reality thanks to the magic of nanotechnology. This little marvel is not just about tiny things; it's about big changes—transformative shifts in how we tackle some of the most pressing environmental challenges we face today. So, grab a cup of coffee, kick back, and let's dive into the ways nanotechnology is reshaping our planet for the better.

First up, let's chat about water purification. You know, water is life. It's that simple. Yet, millions of folks around the globe still struggle to access clean drinking water. That's where nanotechnology struts in, cape billowing in the wind. Researchers are developing nanomaterials that can effectively filter out contaminants, pathogens, and even heavy metals from water. Think of it like having a super sponge that can absorb all the bad stuff while letting the good stuff flow through.

Take, for example, graphene oxide membranes. These nifty little sheets can separate salt from water, making seawater drinkable. Imagine turning the vast oceans into a source of fresh water. That's not just a pipe dream; it's a possibility! And let's not forget about nanoparticles that can destroy harmful bacteria and viruses. They act like tiny knights in shining armor, protecting our health one drop at a time. The implications here are huge, especially in developing countries where clean water is a luxury.

Now, let's pivot to pollution control and remediation. We all know that pollution is a nasty beast, lurking in our air, water, and soil. But fear not! Nanotechnology is here to help tame that beast. One exciting application is the use of nanomaterials in capturing and breaking down pollutants. Picture this: tiny particles that can absorb toxic chemicals from industrial waste or even from the air we breathe. It's like having a clean-up crew that works tirelessly to restore our environment.

Take carbon nanotubes, for instance. These little wonders can capture carbon dioxide from the atmosphere and even convert it into useful products. It's like giving pollution a second chance! And then there are nanomaterials designed for soil remediation. They can break down harmful substances in the ground, making it safe for plants to grow again. This is not just science fiction; it's happening now, and it's making a difference.

But wait, there's more! Let's talk about sustainable materials and energy solutions. We're living in a time when sustainability isn't just a buzzword; it's a necessity. Nanotechnology is leading the charge here, too. By creating new materials at the nanoscale, we can develop products that are not only lighter and stronger but also more eco-friendly.

For example, nanocomposites made from biodegradable materials can replace traditional plastics, reducing waste and pollution. It's like swapping out your old clunker for a shiny new electric car—better for you and the planet! Plus, nanotechnology is revolutionizing energy solutions. Solar panels enhanced with nanomaterials can capture more sunlight and convert it into energy more efficiently. Imagine powering your home with the sun, all thanks to these tiny advancements.

Now, you might be wondering how all this fits together. Well, think of it like a symphony. Each application of nanotechnology plays its part, contributing to a larger, harmonious solution to our environmental woes. Water purification, pollution control, and sustainable materials are all interconnected. When we improve one area, it positively impacts the others. It's a beautiful cycle of improvement that can lead us to a cleaner, greener future.

But let's not get ahead of ourselves. While the potential is enormous, there are challenges to overcome. We need to ensure that these technologies are safe and effective. That means rigorous testing and a commitment to responsible innovation. We can't just throw nanomaterials into the mix without understanding their impact on human health and the environment. It's like baking a cake; you can't skip the steps or the cake will flop.

So, what can you do? Well, it starts with awareness. Educate yourself about nanotechnology and its applications. Share what you learn with others. The more we talk about it, the more we can push for responsible use and development. And if you're feeling particularly inspired, consider a career in this field. There's a world of opportunity waiting for you, and your contributions could be the key to unlocking even greater advancements.

In closing, let's take a moment to visualize the future. Picture a world where every drop of water is pure, where our air is clean, and where our materials are sustainable. It's not just a dream; it's a goal we can achieve together. With the power of nanotechnology, we have the tools to make a real difference. So, roll up your sleeves and get ready to be part of this incredible journey. The future is bright, and it's yours for the taking!

Chapter 6

Nanotechnology in Electronics

In the world of electronics, size matters—more than you might think. The miniaturization of electronic components has transformed the landscape of technology, turning bulky devices into sleek, portable wonders. Picture this: once upon a time, computers took up entire rooms, and now, they fit snugly in our pockets. That's the magic of nanotechnology at work! By manipulating materials at the nanoscale, we're not just shrinking devices; we're enhancing their performance and capabilities in ways that were once the stuff of science fiction.

Let's dive into the heart of this revolution. Miniaturization isn't just about making things smaller; it's about making them smarter. When we talk about electronic components at the nanoscale, we're referring to transistors, capacitors, and other elements that have been shrunk down to sizes so tiny they can't even be seen with the naked eye. Imagine a grain of salt—now imagine a thousand tiny grains, each one packed with power and potential. That's what nanotechnology does. It allows us to create more efficient, faster, and more powerful devices by cramming more functionality into less space.

Now, let's not overlook the impact this has had on computing and communication technologies. With these miniature components, we've seen a dramatic increase in processing speeds and data storage capabilities. Remember the days of dial-up internet? Now, with nanotechnology, we're zipping around the web at lightning speed, streaming movies, and connecting with people across the globe in real-time. It's like we've opened the floodgates to a digital universe, all thanks to the advancements in nanotechnology.

But wait, there's more! The future trends in nanotechnology-driven electronics are nothing short of thrilling. We're on the cusp of breakthroughs that will redefine how we interact with technology. Imagine flexible electronics that can bend and stretch, fitting seamlessly into our lives. Think about smart devices that can monitor our health, adjust our environments, and even learn from our behaviors. The possibilities are endless, and they're all rooted in the innovative spirit of nanotechnology.

As we look ahead, one thing is clear: the integration of nanotechnology into electronics isn't just a trend; it's a paradigm shift. It's a movement toward smarter, more efficient, and more sustainable technology. This isn't just about making our gadgets cooler; it's about enhancing our lives and solving real-world problems.

So, what does this mean for you? It means that you're part of an incredible journey. Whether you're a tech enthusiast, a budding inventor, or just someone curious about the future, there's a place for you in this nanotechnology revolution. Embrace it! Explore how these advancements can improve your life and the lives of those around you. Remember, every great change starts with a single step, and you have the power to be part of this transformation.

The next time you pick up your smartphone or turn on your laptop, take a moment to appreciate the wonders of nanotechnology that make it all possible. It's not just technology; it's a testament to human ingenuity and the endless possibilities that lie ahead. So let's keep pushing the boundaries, dreaming big, and working together to shape a future that's not just smart, but truly extraordinary. You got this!

Chapter 7

The Role of Nanotechnology in Energy

Picture this: a world where energy flows as easily as a summer breeze, where flicking a switch brings light without guilt over harming our planet. Sounds dreamy, right? Well, that dream's not so far off anymore. Thanks to nanotechnology, we're witnessing a revolution in how we think about and use energy. So, strap in! We're diving deep into the ways nanotech is reshaping the energy scene.

Let's kick it off with renewable energy sources. You know, those shiny alternatives that promise to keep our planet spinning without the nasty side effects of fossil fuels. Nanotechnology's stepping into the ring like a heavyweight champ, bringing innovations that ramp up the efficiency of solar panels, wind turbines, and even biofuels.

Take solar energy, for example. We've all seen those big, shiny panels on rooftops, but did you know that nanomaterials can make them even better? Researchers are using nanoparticles to create solar cells that can soak up more sunlight and convert it into electricity. Imagine giving your old car a turbocharger—suddenly, it's zooming down the road, and you're grinning from ear to ear. That's what's happening with solar panels! These advancements mean solar energy can be harvested more efficiently, making it a solid option for homes and businesses. Picture a future where solar panels aren't just efficient but also cheaper and more accessible. That's the magic of nanotechnology!

Now, let's not forget about wind energy. Those towering turbines are impressive, but they could always use a little boost, right? Here comes nanotechnology, helping to create lighter and stronger materials for turbine blades. This means they can catch more wind without toppling over like a house of cards in a storm. Stronger materials mean longer-lasting turbines, which means more energy produced over time. Win-win, folks!

And what about biofuels? Nanotechnology's shaking things up here too. By using nanocatalysts, researchers are speeding up the process of turning biomass into usable energy. Think of it like adding a secret ingredient to your grandma's famous recipe that makes it even tastier. This innovation not only boosts the yield of biofuels but also makes the process more sustainable. So, next time you fill up your tank, you might just be fueling your ride with a little help from the nanotech wizards!

Now, let's switch gears and talk about energy efficiency. Generating energy is one thing, but using it wisely? That's where nanotechnology really shines, like a diamond in the rough. By weaving nanomaterials into our energy systems, we can cut down on waste and amp up performance.

Take insulation, for instance. Traditional insulation can only do so much, but with nanotech, we're talking about materials that reflect heat and keep your home cozy in the winter while staying cool in the summer. It's like wrapping your house in a warm hug, keeping all that precious energy where it belongs—inside. Not only does this save you cash on your energy bill, but it also shrinks your carbon footprint. Who wouldn't want that?

And let's not overlook smart grids. These high-tech marvels are revolutionizing how we distribute and consume energy. By using nanotechnology, we can create sensors and devices that monitor energy use in real-time. It's like having a personal energy coach, nudging you to turn off that light when you leave the room or reminding you to unplug devices that are just sucking up power. With smarter energy consumption, we can maximize efficiency and minimize waste. It's a beautiful dance of technology and sustainability!

Now, let's get to the juicy part—innovations in battery and fuel cell technology. You might be thinking, "What's the big deal? Batteries are batteries, right?" Oh, if only it were that simple! The battery world's evolving faster than a speeding bullet, and nanotechnology is leading the charge.

Lithium-ion batteries have become the gold standard for everything from smartphones to electric vehicles, but they've got their limitations. That's where nanotechnology swoops in like a superhero. By incorporating nanomaterials, we can crank up the surface area of the battery's electrodes, allowing for faster charging and discharging. It's like going from a slow, sleepy turtle to a cheetah on the run! Your phone charges in a snap, and your electric car can zip down the highway without fretting about running out of juice.

And let's not forget about fuel cells. These beauties are the unsung heroes of clean energy, converting hydrogen and oxygen into electricity with water as the only byproduct. But they've faced challenges—mainly cost and efficiency. Thanks to nanotech, we're seeing breakthroughs that make fuel cells more affordable and effective. Researchers are developing catalysts made from nanoparticles that require less precious metals, which can change the game for the industry. Imagine a future where fuel cells power everything from cars to homes, all while being kinder to Mother Earth. That's the dream!

So, what's the takeaway from all this? Nanotechnology isn't just a fancy term tossed around in science labs; it's a powerful tool reshaping our energy landscape. From enhancing renewable energy sources to improving energy efficiency and revolutionizing battery and fuel cell tech, the impact is huge. You've got the power to be part of this movement. Embrace the possibilities, educate yourself, and spread the word. The future of energy is bright, and with nanotechnology, we're lighting the way!

As you continue your journey through the world of nanotechnology, remember this: every small step you take in understanding and advocating for these innovations contributes to a larger change. You're not just a spectator;

you're a vital player in this unfolding story. Let your curiosity guide you, and don't shy away from exploring how you can make a difference. Together, we can harness the power of nanotechnology to create a sustainable and energy-efficient future for all. Now, let's keep that momentum going!

Alright, let's dig deeper into some specific applications of nanotechnology in energy. We've skimmed the surface, but there's so much more to uncover.

First off, let's chat about solar cells a bit more. You might've heard about perovskite solar cells. They're the new kids on the block, and they're turning heads. These cells are cheaper to produce and can be made flexible, which means they can be integrated into a wider variety of surfaces—think windows, walls, or even your clothes! How wild is that? Nanotechnology plays a crucial role in enhancing their efficiency and stability. Researchers are mixing in nanomaterials to improve light absorption and reduce the cost of production. So, the more we learn about these tiny materials, the more we can push solar tech into everyday life.

And speaking of everyday life, let's talk about batteries again. You've probably heard of graphene, right? This wonder material is just one atom thick, yet it's incredibly strong and conducts electricity like nobody's business. Researchers are exploring how to use graphene in batteries to boost their performance. Imagine a battery that charges in minutes instead of hours and lasts longer too! That's not just a pipe dream; it's on the horizon, thanks to nanotechnology.

Now, let's take a moment to appreciate the role of nanotechnology in energy storage. Energy storage is crucial for balancing supply and demand, especially with renewable sources that can be a bit unpredictable. Nanotechnology is helping to create better supercapacitors— devices that store energy and release it quickly. These can charge and discharge much faster than traditional batteries, making them perfect for applications where quick bursts of energy are needed, like in electric vehicles or grid stabilization.

And don't sleep on the potential of nanotechnology in hydrogen production! Hydrogen's got a reputation as a clean fuel, but producing it efficiently has been a challenge. Nanotechnology is stepping in to help with that too. Researchers are developing catalysts made from nanoparticles that can significantly lower the energy needed for hydrogen production through electrolysis. This means we could produce hydrogen more sustainably and at a lower cost, making it a more viable option for clean energy.

Let's not forget about the environmental impact. Nanotechnology isn't just about making energy cleaner; it's also about making the processes that produce energy less harmful. For instance, nanomaterials can be used to create more efficient filters for capturing pollutants from power plants. This means that as we generate energy, we can also minimize the negative impact on the environment. It's like having your cake and eating it too!

And here's a fun thought: what if we could use nanotechnology to make our buildings energy-neutral? Imagine a skyscraper that generates as much energy as it consumes. It sounds like something out of a sci-fi movie, but with advances in nanotechnology, it could become a reality. By integrating nanomaterials into construction, we could create buildings that not only insulate better but also produce energy through embedded solar cells or even piezoelectric materials that generate power from movement. The possibilities are endless!

Let's shift gears again and consider the economic side of things. Investing in nanotechnology for energy isn't just about being eco-friendly; it's also about creating jobs and driving innovation. As companies invest in research and development, they're not just improving energy efficiency—they're also creating new industries and job opportunities. Think about all the tech jobs that come from developing and implementing these new technologies. It's a win for the economy and the environment!

Now, I know some folks might be skeptical. "Isn't nanotechnology just another buzzword?" they might ask. And yeah, it's easy to roll your eyes at the latest trend. But the reality is that the advancements we're seeing are real and impactful. Just look at how quickly electric vehicles have become mainstream. That's partly due to innovations in battery technology, much of which is driven by nanotechnology. It's not just talk; it's action!

And hey, let's not forget about the role of education in all this. If we want to harness the power of nanotechnology for energy, we need to educate the next generation. Schools should be teaching kids about these innovations and how they can get involved. Whether it's through STEM programs or hands-on projects, we need to inspire young minds to think about energy in new ways. Who knows? The next big breakthrough could come from a high school science fair!

As we wrap up this journey through the world of nanotechnology in energy, let's remember that we're all in this together. Whether you're a scientist, a student, or just someone curious about the future, you have a role to play. The more we understand and advocate for these technologies, the more we can push for a sustainable future.

So, keep your eyes peeled for new developments. Stay curious, ask questions, and don't hesitate to share what you learn. The future of energy is bright, and with nanotechnology leading the charge, we're on the brink of something amazing. Let's keep that momentum going and make sure we're all part of this exciting journey!

Chapter 8

Safety and Ethical Considerations

When we talk about nanotechnology, we're diving into a world of tiny particles that can pack a mighty punch. But hold on a second! With great power comes great responsibility, right? Just like a superhero must navigate their abilities carefully, we too need to consider the potential risks that come with the amazing advancements in nanomaterials. It's a bit like walking a tightrope—you gotta keep your balance to avoid a tumble.

Let's kick things off with the potential risks of nanomaterials. These little guys, while revolutionary, aren't without their concerns. You see, at the nanoscale, materials can behave in ways we don't fully understand yet. Think of it like trying to predict the weather in a new city—you might have a general idea, but you could still get caught in a surprise storm. Some nanomaterials can be toxic, and their small size allows them to enter our bodies and ecosystems more easily than larger particles. This raises questions about their long-term effects on health and the environment. It's like opening a box of chocolates—you don't always know what you're gonna get.

Now, let's chat about the regulatory frameworks that are being put in place to manage these risks. Governments and organizations around the world are working hard to create guidelines and regulations that ensure the safe development and use of nanotechnology. It's a bit like building a fence around a playground—keeping the kids safe while they explore and have fun. The U.S. Environmental Protection Agency (EPA) and the Food and Drug Administration (FDA) are just a couple of the key players in this game. They're responsible for assessing the safety of nanomaterials before they hit the market. But here's the kicker: regulations can lag behind innovation. It's a constant game of catch-up, and that's where we need to stay vigilant.

Now, let's pivot a bit and dig into the ethical implications of these advancements. This is where things get really interesting. As we push the boundaries of what's possible with nanotechnology, we must also grapple with the moral questions that arise. For instance, consider the potential for surveillance technologies that use nanomaterials. Imagine tiny sensors that can track your every move— yikes! While this could enhance security, it also raises red flags about privacy. It's a double-edged sword, and we need to wield it wisely.

Moreover, there's the question of access. Who gets to benefit from these advancements? If nanotechnology is only available to the wealthy, we could end up widening the gap between the haves and the have-nots. It's like giving a fancy toy to one kid on the block while the others are left watching from the sidelines. We need to ensure that the benefits of nanotechnology are shared equitably, so everyone can play.

In the grand tapestry of nanotechnology, safety and ethics weave together to create a complex picture. It's not just about what we can do; it's about what we should do. As we move forward, we must keep our eyes wide open, weighing the risks against the rewards. This is a journey we're all on together, and every voice matters.

So, how do we navigate this landscape? Here are a few key takeaways to keep in mind:

1. Stay informed. Knowledge is power. Keep up with the latest research and developments in nanotechnology. Understanding the risks and benefits will empower you to make informed decisions.

2. Advocate for transparency. Whether you're a consumer, a researcher, or a policymaker, push for clear communication about the safety and ethical considerations of nanotechnology. We all deserve to know what's in the products we use.

3. Engage in discussions. Talk about these issues with your friends, family, and colleagues. The more we discuss, the more we can collectively shape the future of nanotechnology.

4. Support responsible innovation. Encourage companies and researchers to prioritize safety and ethics in their work. When we demand accountability, we help steer the industry in a positive direction.

5. Embrace a holistic view. Consider the broader implications of nanotechnology on society, the environment, and our future. It's not just about the science; it's about how it fits into the world we want to create.

Remember, you're not just a passive observer in this journey. You're an active participant, and your voice matters. Together, we can navigate the exciting yet challenging waters of nanotechnology, ensuring that we harness its potential for good while safeguarding our health, our environment, and our ethical standards. So, let's keep the conversation going and step boldly into the future!

Chapter 9

The Future of Nanotechnology Research

As we stand on the brink of a new era in science, the future of nanotechnology research is bursting with promise and potential. It's like watching a flower bloom—every petal unfurling reveals a new layer of beauty and complexity. Emerging fields and interdisciplinary approaches are paving the way for breakthroughs that we can only begin to imagine. You see, the world of nanotechnology isn't just about tiny particles; it's about how those particles can dance with other disciplines, creating a symphony of innovation.

Let's talk about emerging fields first. Imagine the fusion of nanotechnology with fields like biology, materials science, and even art. That's where the magic happens! Researchers are diving into bio-nanotechnology, harnessing the power of nanoscale materials to create smarter, more efficient drug delivery systems. We're talking about tiny robots that can swim through your bloodstream, delivering medicine right where it's needed. Isn't that something? It's like having a personal courier for your health!

Then there's nanotechnology in environmental science. Think about it—tiny particles that can purify water, absorb pollutants, or even capture carbon dioxide from the atmosphere. The possibilities are endless! And as we face global challenges like climate change and resource scarcity, these innovations are not just exciting; they're essential.

Now, let's not forget about the intersection of nanotechnology and artificial intelligence. It's a match made in heaven! AI can help us design nanomaterials with unprecedented precision, predicting how they'll behave in various applications. This collaboration is a game-changer, leading to innovations we've only dreamed of. Imagine smart materials that can adapt to their environment—like a chameleon that changes color to reflect its surroundings!

But here's the kicker: none of this can happen in a vacuum. The future of nanotechnology is being shaped by collaborations between academia and industry. Universities are hotbeds of creativity and research, while industries bring that research to life, transforming ideas into tangible products. It's a beautiful partnership, like peanut butter and jelly—each brings something unique to the table.

Take, for instance, a university working on a new type of nanomaterial that could revolutionize batteries. The academic team conducts groundbreaking research, but it's the industry partners who can scale that research up, turning it into a product that can power our devices. This collaboration is vital, creating a feedback loop where research informs industry needs, and industry challenges inspire new research questions. It's a win-win for everyone involved!

And let's not forget the role of government and funding agencies in this mix. They're like the fuel that keeps this engine running. By investing in nanotechnology research, they're not just supporting scientists; they're investing in the future of our society. When funding flows into interdisciplinary projects, it opens doors to innovation that can change lives. It's like planting seeds in a garden—nurturing them until they grow into something magnificent.

As we look ahead, we can see trends shaping the future of nanotechnology. One of the most exciting is the push for sustainability. Researchers are increasingly focused on developing green nanotechnology—materials and processes that minimize environmental impact. This isn't just about creating cool gadgets; it's about being responsible stewards of our planet. The next generation of nanotechnologists will be armed with the knowledge and tools to create solutions that are not only effective but also sustainable.

Moreover, the rise of personalized medicine is another trend to watch. With nanotechnology, we're moving toward treatments tailored specifically to an individual's genetic makeup. This means more effective therapies with fewer side effects. It's like having a tailor for your health—everything fits just right!

But let's not get too caught up in the future without acknowledging the present. Right now, researchers are making strides in various applications of nanotechnology, from electronics to healthcare to environmental remediation. It's a vibrant field, full of passionate individuals eager to make a difference. And you, dear reader, have a role to play in this journey.

Whether you're a budding scientist, an entrepreneur, or simply someone curious about the world, there's a place for you in the nanotechnology revolution. Engage with the community, ask questions, and never stop learning. The more you understand, the more you can contribute to this exciting field.

So, what can you do today? Start by exploring the latest research articles, attending workshops, or even reaching out to local universities to see how you can get involved.

The future is bright, and it's waiting for you to step in and make your mark.

In summary, the future of nanotechnology research is a thrilling tapestry woven from emerging fields, interdisciplinary approaches, and strong collaborations between academia and industry. As trends continue to evolve, the potential for transformative change is limitless. And remember, every small step you take can lead to monumental advancements. So keep that curiosity alive, and let's embark on this incredible journey together!

As we wrap up this chapter, visualize the impact your work could have. Imagine a world where nanotechnology solves some of our most pressing challenges—clean water, effective healthcare, sustainable energy. Picture yourself at the forefront of this movement, contributing to innovations that change lives. You've got this! Embrace the journey, and let's see where it takes us.

Chapter 10

Nanotechnology and Consumer Products

Imagine walking through your local store, just browsing, and bam! You're surrounded by a sprinkle of magic—yep, that's nanotechnology at work in your everyday products. From the toothpaste you scrub with to the sneakers you rock, this tiny tech is shaking things up in ways you might not even notice. Let's dig into how this little marvel is jazzing up our lives, why you should keep your ear to the ground about it, and what cool stuff is just around the corner.

First off, let's chat about those products you use every day. You probably don't give a second thought to the shampoo you lather in or the sunscreen you slather on before hitting the beach. But guess what? A lot of these goodies are getting a serious facelift thanks to nanotechnology. Seriously! Those nanoparticles—tiny bits of stuff smaller than a cell—are stepping in to boost the performance of your favorite products.

Take sunscreen, for instance. We all know the struggle of looking like a ghost after applying traditional sunscreen. But with nanotech, companies are using super-small particles of zinc oxide or titanium dioxide that practically disappear on your skin. So, you get that sun protection without the chalky aftermath. Pretty neat, right? And that's just scratching the surface.

Now, let's talk about your clothes. You've seen those "stain-resistant" tags, right? Well, guess who's behind that? You guessed it— nanotechnology! Fabrics are treated with nanoparticles that repel water and stains, keeping your favorite shirt looking fresh longer. It's like having a magic shield against spills and splashes. Who wouldn't want that?

And what about your gadgets? Your smartphone, tablet, or laptop is probably loaded with nanotech, making them faster, lighter, and more efficient. Companies are using nanoscale materials to whip up batteries that charge in a flash and last way longer. Imagine never stressing about your phone dying on you when you need it most. That's the kind of future we're headed toward, folks!

But here's the kicker: while all this innovation is super exciting, it's crucial for you to stay in the loop about what's in the stuff you're using. Knowledge is power, right? It's not just about the cool features; it's also about understanding what using nanotech means for our daily lives.

You might be thinking, "How do I stay informed?" Start by checking out labels and doing a little homework on brands that are all about transparency. Companies are waking up to the need for consumer education, and many are stepping up to spill the beans about the nanomaterials they use. Look for resources that break down how these materials work and why they're beneficial.

Chat it up with friends and family about nanotechnology. Share what you learn and get them curious too. The more we talk about it, the more we raise awareness. It's a team effort, and every little bit of knowledge helps create a more informed crowd.

Now, let's look ahead. The future of consumer goods is bursting with potential, and nanotech is leading the charge. Imagine a water bottle that not only keeps your drink cold but also purifies it while you're on the go, all thanks to nanotechnology. Or picture your car magically fixing minor scratches with a special nanomaterial. These aren't just pipe dreams—they could be reality before you know it.

We can't forget about food packaging either. With nanotech, we can whip up materials that keep food fresher for longer, cutting down on waste and ensuring you get the most out of your groceries. It's a win-win for your wallet and the planet.

And let's talk health and beauty. The beauty world is already diving into nanotech to create skincare products that penetrate deeper and deliver results more effectively. Think serums that target specific skin issues at a cellular level. It's like having a spa day in a bottle!

So, what's your move? Stay curious! Keep an eye out for new products that utilize nanotechnology and embrace the innovations coming your way. Be proactive with your choices and support brands that prioritize safety and transparency in their use of nanomaterials.

In the grand scheme of things, integrating nanotechnology into consumer products is just the tip of the iceberg. It's a thrilling time to be alive, and you're right in the thick of it. As you wander through the aisles of your favorite stores, remember that each product is a testament to human creativity and the relentless quest for improvement.

So, next time you grab a bottle of shampoo or a shiny new gadget, take a sec to appreciate the science behind it. You're not just using a product; you're part of a revolution that's reshaping how we live, work, and play.

Embrace it, share it, and most importantly, enjoy the ride!

As we wrap up this little journey through nanotechnology in consumer products, keep this in mind: your awareness and engagement are key to shaping the future. The more you know, the better choices you can make, and that's where the real power lies. So dive into this fascinating world, and let it inspire you to be part of the change. Your adventure into the future starts now!

Now, let's get a bit deeper into the nitty-gritty of nanotechnology and how it's creeping into our lives, shall we? You might be surprised to learn just how many products you encounter daily are getting a nanotech makeover.

Let's kick it off with cosmetics. Ladies and gents, if you've ever applied foundation or moisturizer, you might've unknowingly used nanotech. Many beauty products are now infused with nanoparticles that help them absorb better into the skin. It's all about that smooth finish, right? These tiny particles can also carry active ingredients deeper into the skin, giving you more bang for your buck. It's like having a tiny army of helpers working to make your skin glow!

And speaking of glowing, let's not overlook the world of health supplements. Some companies are leveraging nanotechnology to create supplements that deliver nutrients more effectively. The idea is that smaller particles can be absorbed more easily by the body, meaning you get more of those good-for-you vitamins and minerals. So, if you're popping a multivitamin, there's a chance it's been zapped with some nanotech magic to boost its effectiveness.

Now, let's switch gears to food. Yep, nanotechnology is even making waves in the kitchen! Food packaging is getting a high-tech upgrade. You know those annoying little packets of silica gel that say "Do not eat"? Well, some food packaging now uses nanosensors to detect spoilage. That means your food could stay fresher longer, and you won't have to worry about tossing out expired stuff as often. Plus, it's a step towards reducing food waste, which is a huge win for the environment.

And while we're on the topic of food, let's chat about flavor enhancers. Some food companies are experimenting with nanotechnology to create flavoring agents that can deliver a punch without adding extra calories. Imagine enjoying your favorite snacks without the guilt. Now that's a tasty thought!

But hold on—let's not get too carried away without addressing the elephant in the room.

With all this cool tech, there are some folks raising eyebrows about safety and ethics. Are these nanoparticles safe for us? What about the environment? These are valid concerns. It's crucial for companies to be transparent about their practices and for consumers to stay informed.

So, how do we navigate this brave new world? First off, do your homework. Research brands and products that are upfront about their use of nanotechnology. Check out consumer reviews and expert opinions. Don't be afraid to ask questions. If you're curious about what's in a product, reach out to the company. Most are more than happy to chat about their tech.

And here's a thought: consider supporting companies that prioritize sustainable practices. Many brands are now focusing on eco-friendly materials and processes, so you can feel good about what you're buying. It's all about making choices that align with your values.

Now, let's get back to the fun stuff! The potential for nanotechnology in consumer products is mind-blowing. Picture this: smart textiles that can monitor your health. Imagine wearing a shirt that tracks your heart rate or body temperature and sends that info to your phone. It's like having a personal health assistant right on your back! This kind of innovation could revolutionize how we approach health and wellness.

And what about smart home devices? You know those voice-activated assistants that help you control your lights and thermostat? Well, nanotechnology could take them to the next level. Imagine devices that can learn from your habits and adjust themselves to create the perfect environment in your home. Talk about a cozy living space!

Let's not forget about the automotive industry. Cars are getting smarter and more efficient, thanks to nanotech. We're talking about lightweight materials that improve fuel efficiency and advanced sensors that enhance safety features. And who knows? Maybe in the near future, we'll see cars that can communicate with each other to avoid accidents. It's like something out of a sci-fi movie!

And while we're dreaming big, let's think about the environment. Nanotechnology has the potential to tackle some serious environmental issues. For instance, it can be used in water purification systems, making it easier to access clean drinking water in areas where it's scarce. That's a game changer for public health!

Now, let's get a bit personal. Think about your daily routine. How many products do you use that could benefit from nanotechnology?

Your morning coffee? What if there was a coffee maker that used nanotech to brew the perfect cup every time? Or your workout gear? Imagine leggings that wick away sweat while also providing muscle support. The possibilities are endless!

And here's a little nugget of wisdom: stay open-minded. Technology is always evolving, and while it's easy to be skeptical, it's also exciting to see how it can improve our lives. Embrace the changes and be part of the conversation about how we can use these advancements responsibly.

In closing, let's circle back to the heart of the matter. Nanotechnology is weaving its way into our lives, and it's not slowing down. From cosmetics to electronics, food to fashion, this tiny tech is making a big impact. So, the next time you're in the store, take a moment to think about the science behind the products you use.

You're not just a consumer; you're part of a movement that's shaping the future.

So, what are you waiting for? Get out there, explore, and stay curious! The world of nanotechnology is waiting for you to dive in. Your journey into this fascinating realm starts now, and who knows what you'll discover along the way?

Chapter 11

The Global Impact of Nanotechnology

Nanotechnology isn't just a buzzword floating around in science circles; it's a force that's shaping our world in ways we're only beginning to understand. Picture this: tiny particles, smaller than a speck of dust, wielding the power to transform economies, tackle global challenges, and foster international collaboration. It's like having a secret weapon in our back pocket, ready to take on the biggest issues we face today. Let's dive into how nanotechnology is making waves across the globe.

First off, let's talk about economic development. Nanotechnology has the potential to ignite economic growth like a spark in a dry field. Countries that invest in this field are not just keeping up; they're setting the pace. Think about it—new industries are sprouting up, jobs are being created, and existing sectors are getting a much-needed facelift. From healthcare to energy, nanotechnology is enhancing productivity and efficiency, leading to better products and services. It's like a domino effect; one innovation leads to another, creating a robust economic ecosystem.

Consider the impact on developing nations, where resources can be scarce. With nanotechnology, there's a chance to leapfrog traditional methods and embrace cutting-edge solutions. Water purification systems that use nanomaterials can provide clean drinking water, while agricultural innovations can boost crop yields. This isn't just science fiction; it's happening right now. When countries harness the power of nanotechnology, they're not just improving their economies—they're elevating the quality of life for their citizens.

Now, let's shift gears and tackle some global challenges. Climate change, healthcare access, food security—these aren't just headlines; they're real issues that affect millions of lives. Nanotechnology steps in like a superhero, offering solutions that are both innovative and effective. For instance, nanomaterials can be used to create more efficient solar panels, making renewable energy more accessible. In medicine, targeted drug delivery systems can revolutionize how we treat diseases, ensuring that patients receive the right medication in the right dosage at the right time.

And food security? You bet! Nanotechnology can enhance the nutritional content of crops and help them resist diseases, which is crucial as the global population continues to rise. Imagine a world where hunger is a thing of the past, all thanks to tiny particles working their magic. That's the kind of future we're talking about here.

But none of this can happen in isolation. International collaboration is key to unlocking the full potential of nanotechnology. Countries around the globe are coming together, pooling their resources, knowledge, and expertise. Research initiatives are sprouting up like wildflowers in spring, creating networks of scientists, engineers, and innovators who are all working toward a common goal.

Take, for example, the Global Nanotechnology Initiative. This collaborative effort brings together researchers from various countries to share findings, develop standards, and promote safe practices in nanotechnology. It's a beautiful thing, really—when nations set aside their differences and work together for the greater good.

And let's not forget about the role of universities and private sector partnerships. They're like the backbone of this movement, driving research and development forward. When academia and industry join forces, magic happens. Breakthroughs that once seemed impossible become reality, and the benefits ripple out to communities worldwide.

Now, I know this might sound overwhelming. The challenges we face are immense, and the solutions aren't always straightforward. But here's the thing: with every challenge comes an opportunity. Nanotechnology is a beacon of hope, a tool we can wield to create positive change. It's about harnessing our collective creativity and ingenuity to tackle the issues that matter most.

So, what can you do? First, stay informed. Knowledge is power, and understanding the impact of nanotechnology can empower you to engage in conversations about its potential. Second, consider how you can contribute. Whether it's through your career, volunteering, or simply sharing information with others, every little bit helps.

As we look to the future, let's embrace the possibilities that nanotechnology offers. Picture a world where clean water, efficient energy, and accessible healthcare are the norm. It's not just a dream; it's within our reach. Together, through collaboration and innovation, we can create a brighter future for all.

Remember, the journey of a thousand miles begins with a single step. So let's take that step together, armed with the knowledge that nanotechnology is more than just a field of study—it's a pathway to a better world. Keep pushing forward, stay curious, and never underestimate the power of tiny particles to create monumental change. You've got this!

Chapter 12

Nanotechnology in Agriculture

Picture this: fields sprawling out like a vibrant patchwork quilt, greens and golds dancing in the breeze. Farmers, the unsung heroes of our society, work their tails off to feed us all. But with climate change, a booming population, and dwindling resources knocking at our door, we need some serious innovation. That's where nanotechnology steps in—think of it as the superhero of agriculture, swooping in to save the day.

So, what's the deal with nanotechnology? Imagine crops that not only sprout faster but also fend off diseases that used to wipe out entire harvests. Sounds like magic, right? Well, it's not. It's all about manipulating materials at the nanoscale. Scientists are whipping up smarter fertilizers and pesticides that hit the mark with laser-like precision. This means healthier plants, less chemical runoff, and a food supply that's looking brighter than ever. It's like giving Mother Nature a turbo boost—farmers' secret weapon against the wild cards of weather and pests.

But hold on, there's more to this story! Sustainable farming practices are the beating heart of this revolution. With the global population set to hit nearly 10 billion by 2050, the heat is on for our food systems. Old-school farming often leads to soil depletion and a whole lotta water waste. Enter nanotechnology, ready to save the day! These tiny materials can jazz up soil quality, help retain water, and make it easier for plants to soak up nutrients. It's like giving Mother Earth a rejuvenating spa day, ensuring she stays fertile and productive for generations to come.

Now, let's visualize a farmer in the Midwest, armed with seeds that have been supercharged by nanotechnology. These seeds sprout quicker and stronger, giving the farmer crops that can take on droughts and diseases while using fewer resources. Talk about a win-win! Higher yields? Check. Smaller environmental footprint? Double check. This is what sustainable farming is all about—feeding the world without putting our planet in a chokehold.

And hey, food security is a biggie on everyone's radar these days. The future of food security is tightly woven with the strides we're making in nanotechnology. If we harness these tiny innovations right, we can build a more secure food supply. Picture a world where hunger is a relic of the past, where every kid has access to nutritious meals, and farmers can breathe easy without worrying about crop failures. That's the dream we're chasing.

But how do we get from here to there? It all kicks off with education and awareness. Farmers, researchers, and consumers need to get hip to what nanotechnology can do for agriculture. By teaming up scientists with agri-experts, we can create a knowledge-sharing vibe that lifts everyone involved. This isn't just about tech; it's about creating a community that believes in the power of innovation to change the world for the better.

And here's the kicker: you don't need a lab coat to join the party. Whether you're an aspiring entrepreneur, a curious student, or just someone who cares about the future of food, there's a spot for you in this nanotech revolution. Push for sustainable practices, back local farmers who are diving into these advancements, and keep your ear to the ground about the latest research. Every little action counts!

Looking ahead, the future of agriculture is bursting with potential. With nanotechnology on our side, we can cultivate a world that thrives on innovation and sustainability. Let's celebrate these tiny technologies that pack a colossal punch. Together, we can plant the seeds of change, ensuring that the earth's bounty is within reach for everyone.

So, let's roll up our sleeves and dig in! The road ahead is full of promise. With each step we take, we're not just growing crops; we're nurturing a future where food security and sustainability are best buds. You got this! Embrace the possibilities, and let's turn this vision into reality.

Now, let's dive a bit deeper into the nitty-gritty of nanotechnology in agriculture. We're talking about how these innovations are not just cool science experiments but real-life applications that can change the game. For instance, the use of nanoparticles in fertilizers can lead to more efficient nutrient delivery. Traditional fertilizers often end up leaching away, wasting resources and polluting waterways. But with nanotechnology, nutrients can be encapsulated in tiny particles that release them slowly, giving plants exactly what they need when they need it. It's like a slow cooker for crops—no more overcooked veggies!

And then there's the world of pest control. Conventional pesticides can be a double-edged sword. They're effective at keeping pests at bay, but they also wreak havoc on beneficial insects and the environment. Nanotechnology offers a more targeted approach. Imagine using nanoscale pesticides that only affect the pests and leave the good guys alone. It's a win for farmers and the ecosystem alike. Plus, with less chemical use, we're looking at healthier food options for consumers. Win-win-win!

Let's not forget about water management. Water scarcity is a huge concern, especially in farming regions. Nanotechnology can help here too. Nanomaterials can improve water retention in soil, making it easier for plants to access moisture. Some studies even suggest that nanotechnology can be used in irrigation systems to optimize water usage. It's like giving plants a personal hydration system, ensuring they stay quenched without wasting a drop.

Now, let's get a bit personal. Picture your favorite farmer—maybe it's your uncle or a neighbor down the road. They've been doing things the old-fashioned way for years, and now they're faced with the pressures of modern agriculture. With climate change throwing curveballs and prices fluctuating, it's a tough gig. But imagine if they could adopt nanotechnology. They could boost their yields, cut down on waste, and ultimately make their operation more profitable. It's not just about feeding the world; it's about empowering farmers to thrive in a challenging landscape.

But, hey, it's not all sunshine and rainbows. There are challenges to consider. The adoption of nanotechnology in agriculture comes with its own set of hurdles. For starters, there's the cost. Not every farmer can afford the latest tech, and that's a real concern. It's crucial to find ways to make these innovations accessible to everyone, not just the big players.

Then there's the regulatory landscape. New technologies often face scrutiny, and rightly so. We need to ensure that these innovations are safe for both humans and the environment. It's a balancing act—embracing progress while keeping safety at the forefront.

And let's talk about education again. Farmers need training to implement these technologies effectively. It's not enough to just throw some fancy tech their way. They need to understand how to use it, what benefits it brings, and how to integrate it into their existing practices. That's where community support comes in—workshops, seminars, and local initiatives can bridge that gap.

Now, let's shift gears and talk about the consumer side of things. You and I, we play a role in this whole equation too. As consumers, we can advocate for sustainable practices and support farmers who are adopting nanotechnology. When you buy from local farmers or choose products that use innovative practices, you're sending a message. You're saying you care about where your food comes from and how it's produced.

Plus, staying informed is key. The more we know about these advancements, the better equipped we are to make choices that align with our values. So, read up on nanotechnology in agriculture, follow the latest news, and don't be afraid to ask questions. Knowledge is power, my friend.

As we dig deeper into this topic, it's clear that the potential of nanotechnology in agriculture is massive. It's not just about boosting yields or cutting down on waste; it's about creating a sustainable future for our food systems. It's about ensuring that future generations have access to nutritious food without compromising the health of our planet.

So, what's next? The future of agriculture is ripe for innovation. We're standing at the crossroads of tradition and technology, and it's up to us to steer the ship. With the right support, education, and collaboration, we can create a world where farmers thrive, food is abundant, and the environment flourishes.

Let's not forget the role of policy-makers. They need to get in on the action too. Supporting research and development in nanotechnology, providing funding for farmers to adopt these practices, and creating a regulatory framework that encourages innovation while keeping safety in mind—these are all crucial steps.

In conclusion, the landscape of agriculture is changing, and nanotechnology is leading the charge. It's a thrilling time to be involved in this field, whether you're a farmer, a consumer, or just someone who cares about the future of food. The possibilities are endless, and the potential for positive change is huge.

So, let's roll up our sleeves and get to work! The journey ahead is filled with promise, and with each step we take, we're not just planting seeds; we're sowing the future of food security and sustainability. Let's embrace the possibilities and turn this vision into reality. You in? Let's do this!

Chapter 13

The Intersection of Nanotechnology and Artificial Intelligence

Now, let's dive into a thrilling world where nanotechnology meets artificial intelligence. Picture this: tiny materials, smaller than a speck of dust, working hand in hand with smart algorithms that can think, learn, and adapt. It's like a dance between two giants of innovation, and the possibilities are downright electrifying!

First off, let's talk about AI-driven innovations in nanotechnology research. Imagine scientists, armed with AI tools, sifting through mountains of data faster than you can say "nanoparticle." These AI systems can analyze complex patterns and predict outcomes with incredible accuracy. They're not just crunching numbers; they're revolutionizing how we understand and manipulate materials at the nanoscale. It's like having a super-smart buddy who knows exactly what you need before you even ask!

Take, for instance, the development of new nanomaterials. AI can assist researchers in discovering novel compounds that could have never been identified through traditional methods. It's akin to having a treasure map that leads you to hidden gems in the vast ocean of scientific knowledge. By leveraging machine learning algorithms, researchers can predict how these materials will behave in various environments, making the process faster and more efficient. This collaboration is not just a passing trend; it's setting the stage for groundbreaking advancements that could redefine entire industries.

Now, let's shift gears and explore the applications of AI in nanomaterial design. When you think about designing a new material, it's not just about mixing elements and hoping for the best. It's a meticulous process, where every atom counts. Here's where AI comes in, acting as a master architect, helping to design materials that are not only effective but also sustainable.

Imagine you're building a house. You wouldn't just throw bricks together without a blueprint, right? Similarly, AI creates sophisticated models that predict how different nanomaterials will interact with one another. This ensures that the final product is optimized for performance, durability, and safety. For example, in medicine, AI can help design nanoparticles that can target specific cells in the body, improving drug delivery systems and reducing side effects. This is where the magic happens—combining the precision of nanotechnology with the intelligence of AI creates a powerful synergy that can lead to life-changing innovations.

Now, let's peek into the future and consider the implications of this intersection. The possibilities are as vast as the universe itself! As AI continues to evolve, we'll see even more sophisticated applications in nanotechnology. Think about smart materials that can adapt to their environment, like a chameleon changing colors. Or consider nanobots that can diagnose diseases at an early stage and deliver treatment precisely where it's needed. This isn't science fiction; it's the future we're stepping into!

But hold on—this future isn't without its challenges. As we harness the power of AI and nanotechnology, we must also navigate the ethical and safety concerns that come along for the ride. We've got to ensure that these technologies are used responsibly and for the betterment of humanity. It's like driving a powerful car; you've got to know how to handle it safely.

So, what does this all mean for you? Well, you're not just a spectator in this grand performance. You have a role to play in this exciting journey! Whether you're a student, a professional, or simply someone curious about the world, there are countless opportunities to engage with these technologies.

Start by educating yourself—read books, attend workshops, and connect with like-minded individuals. You can even get your hands dirty with some DIY nanotechnology projects. The more you learn, the more equipped you'll be to contribute to this revolution. Remember, every small step you take can lead to monumental changes down the road.

In conclusion, the intersection of nanotechnology and artificial intelligence is a realm bursting with potential. It's a partnership that promises to reshape our world in ways we can only begin to imagine. So, keep your eyes wide open and your mind curious. Embrace the journey ahead, and who knows? You might just find yourself at the forefront of the next big breakthrough!

Now, take a moment to visualize your success in this field. Picture yourself contributing to a project that changes lives or creating a solution that makes the world a better place. That vision is not just a dream; it's a possibility waiting for you to seize it. So, roll up your sleeves and get ready to dive into this exhilarating adventure. The future is bright, and it's yours for the taking!

Chapter 14

Case Studies in Nanotechnology

In the realm of nanotechnology, it's not just about the science—it's about the stories that unfold. Picture this: you're sitting back with a warm cup of coffee, ready to explore some fascinating case studies that highlight the incredible applications of nanotech across various industries. These stories not only showcase success but also reveal lessons learned and insights from the pioneers who've been trailblazing this path. So, let's roll up our sleeves and dive in!

Let's kick things off with some jaw-dropping success stories. You might not realize it, but nanotechnology has made a splash in fields you'd never expect. Take medicine, for example. Researchers have whipped up nanoparticles that can zero in on cancer cells with laser-like precision. Imagine a tiny soldier, cruising through your bloodstream, hunting down those rogue cancer cells while leaving healthy ones alone. It's like a targeted drug delivery system straight out of a sci-fi flick. This isn't just some futuristic dream; it's happening right now, saving lives and reshaping how we think about treatment.

And then there's the electronics scene. We've all seen our gadgets shrink in size while packing a punch. That's nanotech doing its thing! From lightning-fast processors to super-efficient batteries, the magic of manipulating materials at the nanoscale has flipped our daily lives upside down. Think about your smartphone for a sec. It's a little engineering marvel, all thanks to the innovations that nanotechnology has sparked.

But wait—there's more! Let's not forget the environmental side of things. Nanotechnology is stepping up to tackle some of our biggest headaches, like water purification. Picture this: tiny nanomaterials filtering out nasty pollutants from water sources, making clean drinking water more accessible. This is a game-changer for communities worldwide. It proves that nanotech isn't just about high-tech toys; it's about making a real difference in people's lives.

Now, while those successes are impressive, the real gold lies in the lessons learned from these projects. One major takeaway? Collaboration is key. Many of the most successful nanotech projects sprang from partnerships between academia, industry, and government. When experts from different fields join forces, it's like a potluck dinner—everyone brings their best dish, and together, they whip up something extraordinary.

Another lesson? Don't skip on rigorous testing and safety protocols. We've seen that the potential risks tied to nanomaterials can't just be brushed aside. Projects that made safety a priority from the get-go have had way more success in gaining public trust and regulatory approval. It's a solid reminder that in our rush to innovate, we need to tread carefully, ensuring our creations are safe for folks and the planet alike.

Now, let's shine a spotlight on some of the pioneers in this field. These trailblazers haven't just pushed the boundaries of science; they've also shared their wisdom to inspire the next wave of innovators. Take Dr. Chad Mirkin, for instance. He's a big name in the development of nanostructures, and his work has revolutionized everything from medicine to electronics. His mantra? Embrace curiosity. He often says that the best discoveries come from asking the right questions and being open to exploring the unknown. That's a powerful nugget for all of us—stay curious and let your imagination run wild.

Then there's Dr. Angela Belcher. She's been using nanotech to whip up sustainable materials for energy storage. Her approach blends biology with engineering, showing us that nature often holds the keys to innovative solutions. She encourages aspiring scientists to think outside the box and consider how mixing disciplines can lead to breakthroughs. It's a call to action for everyone to embrace diverse perspectives and collaborate across fields.

Reflecting on these case studies, it's clear that nanotechnology's impact is profound and far-reaching. Each success story is a testament to human ingenuity and the relentless pursuit of knowledge. But more than that, they remind us that the journey is just as crucial as the destination. The challenges faced, the lessons learned, and the insights gained all contribute to the rich tapestry of this field.

So, what can you take away from all this? First off, recognize that you've got the power to be part of this thrilling revolution. Whether you're a student, a professional, or just a curious soul, there are tons of opportunities for you to engage with nanotechnology. Jump into research, hit up workshops, or even just strike up conversations with experts in the field. Your unique perspective and passion could lead to the next big breakthrough.

Second, don't shy away from collaboration. Remember that potluck analogy? Bring your ideas to the table and invite others to share theirs. You never know what kind of magic can happen when different minds come together. The world of nanotech thrives on teamwork, and your contributions matter.

And finally, stay curious and keep learning. The nanotechnology landscape is always shifting, and there's always something new to uncover. Embrace the journey and don't be afraid to ask questions. Your curiosity could lead you down paths you never even dreamed of.

In wrapping this up, the case studies in nanotechnology are more than just tales of success; they're a celebration of human potential. They remind us that through collaboration, innovation, and a commitment to safety, we can craft a brighter future for ourselves and generations to come. So go ahead—take that first step, and let your journey in the world of nanotechnology kick off. The possibilities are endless, and who knows? You might just be the next pioneer in this exciting field.

Now, let's dig a little deeper. Think about the practical implications of nanotechnology. You know, it's not just about flashy gadgets or cutting-edge medical treatments. It's about the everyday stuff, too. For example, consider the food industry. Nanotech is making waves in food packaging. Imagine a bag of chips that can tell you when it's gone stale. Yep, that's happening! By embedding nanosensors in packaging, companies can enhance food safety and reduce waste. It's like having a personal food critic right in your pantry!

And how about textiles? Nanotechnology is revolutionizing the way we think about clothing. We're talking about fabrics that repel stains, resist wrinkles, and even regulate temperature. Picture a shirt that keeps you cool in the summer and warm in the winter—sounds like magic, right? Well, it's all thanks to nanotech. It's changing the game for fashion and functionality.

But let's not forget the energy sector. Nanotechnology is playing a crucial role in renewable energy solutions. Solar panels, for instance, are becoming more efficient thanks to nanomaterials that enhance light absorption. This means more energy from the sun can be converted into usable power. It's a win-win for the environment and our wallets.

Now, I know what you're thinking. "This all sounds great, but what about the risks?" And you're right to ask! With any new technology, there are concerns. The potential health and environmental impacts of nanomaterials are still being studied. We need to be vigilant. But here's the thing: responsible innovation is key. By prioritizing safety and transparency, we can mitigate risks while reaping the benefits.

So, how do we ensure that nanotechnology develops responsibly? It starts with education. We need to spread awareness about what nanotech is and how it can impact our lives. Schools, universities, and organizations should be at the forefront of this effort. The more people understand nanotechnology, the more informed decisions they can make about its use.

And let's not forget about policy. Governments need to step up and create regulations that keep pace with advancements in nanotech. It's all about finding that sweet spot between fostering innovation and protecting public health. We can't let the excitement of new tech overshadow our responsibility to society.

Now, let's take a moment to think about the future. Where's nanotechnology headed? The possibilities are mind-boggling! We could see breakthroughs in personalized medicine, where treatments are tailored to individual patients at the molecular level. Imagine a world where diseases are not just treated but prevented before they even start. That's the kind of future nanotech could help us build.

And what about smart cities? Nanotechnology could play a crucial role in developing infrastructure that's not just efficient but also sustainable. From smart grids to self-repairing materials, the urban landscape could be transformed. It's about creating environments that work for us and the planet.

But let's keep it real. The journey won't be without its bumps. There'll be challenges, debates, and ethical dilemmas to navigate. But that's part of the beauty of progress. It's messy, it's complicated, but it's also incredibly rewarding.

So, as we wrap this up, let's keep the conversation going. Share your thoughts, ask questions, and engage with others about nanotechnology. Whether it's at a coffee shop, a community center, or even online, every discussion counts. Who knows? You might spark the next big idea or inspire someone else to join the movement.

In the end, the case studies in nanotechnology aren't just about the tech itself; they're about us—our creativity, our curiosity, and our commitment to making the world a better place. So let's embrace this journey together, keep pushing boundaries, and see where it takes us. The future is bright, and it's ours for the taking!

Chapter 15

Your Role in the Nanotechnology Revolution

So, picture this: you're standing on the edge of something huge. A wave of innovation is crashing all around you, and guess what? You're not just watching from the sidelines; you're a key player in this game. Yup, nanotechnology isn't just for lab-coat-wearing scientists or bigwig corporate types. It's for you—regular folks with dreams, curiosity, and a drive to make a difference. Let's jump in and see how you can get involved, find your spot, and help ensure this tech is used responsibly.

First things first, how do you even get into nanotechnology? Well, let's break it down nice and simple. Start with education. No need to be a genius or have a fancy degree to get your feet wet. There's a treasure trove of resources out there just waiting for you. Online courses, community college classes, and workshops are popping up like mushrooms after rain. You can dive into the basics of nanotech, explore its applications, and even get a grip on the ethical issues it raises. Don't forget about local libraries and community centers—they often host talks and events that can ignite your interest and help you meet folks who think like you.

Feeling a bit bolder? Why not chase a career in this field? The opportunities are as endless as the ocean! You could dive into research and development, regulatory affairs, marketing, or even education. There's a niche for everyone. Maybe you're drawn to healthcare, working on groundbreaking therapies, or perhaps environmental science, crafting sustainable solutions for our planet's pressing issues. The beauty of nanotechnology? It's a melting pot of disciplines. So whether you're a tech geek, a creative thinker, or a natural problem-solver, there's a spot for you.

But here's the kicker: as you step into the world of nanotech, keep in mind the balance between innovation and responsibility. This field can change lives, but with that power comes a hefty dose of responsibility. It's super important to approach your work with ethics and sustainability in mind. Think about how your contributions can make a positive impact on society and the environment. Ask yourself, "How can I innovate while keeping an eye on the consequences?" This mindset not only sets you apart but also helps build a culture of responsibility in the field.

Now, let's take a moment to visualize your journey. Picture yourself at a workshop,

surrounded by people who share your passion.
You're learning, exchanging ideas, and feeling
that electric buzz of inspiration in the air. Now,
imagine landing a job where you're not just
clocking in and out, but actually making a
difference. Maybe you're working on a new
nanomaterial that could change the game for
clean energy. Or perhaps you're part of a team
creating a drug delivery system that targets
cancer cells better than ever. It's not just a job;
it's a calling!

If you're still scratching your head about
where to start, here's a little exercise: jot down
your interests and skills. What gets you fired
up? Are you a numbers whiz, or do you thrive
in creative spaces? Once you've got that list, dig
into how those skills can mesh with
nanotechnology. Love writing? Why not check
out science communication? Great at
organizing? Project management in a nanotech
firm could be your jam! The trick is finding that

sweet spot where your passions meet the needs of the field.

As you embark on this journey, don't forget about the power of community. Surround yourself with others who are just as pumped about nanotech as you are. Join online forums, hit up local meetups, or dive into hackathons. These connections can lead to collaborations, mentorships, and friendships that might last a lifetime. Remember, you're not in this alone; there's a whole world of people out there ready to support you.

And speaking of support, let's chat about innovation. The world of nanotechnology is always shifting, which means there's always room for fresh ideas. Don't shy away from thinking outside the box. Challenge the status quo! What problems do you see in your community that could use a nanotech solution? Maybe it's improving water quality or creating more efficient solar panels. Your unique perspective could spark the next big breakthrough. Embrace that creativity and let it fuel your passion.

Now, as you engage with nanotechnology, keep in mind the importance of responsible use. This is a powerful tool, and it's on you to wield it wisely. Advocate for ethical practices in your workplace and beyond. Stay up to date on the latest regulations and safety standards. Engage in discussions about the societal impacts of nanotechnology. By being an informed and responsible participant, you can help shape the future of this field for the better.

Let's wrap this up with a little pep talk. You have the power to make a difference in the nanotechnology revolution. Your voice matters, your ideas matter, and your actions matter. Don't let fear or doubt hold you back. Embrace the challenges as opportunities for growth.

Remember, every small step you take brings you closer to your goals. So get out there, educate yourself, connect with others, and innovate with purpose. The future is bright, and it's waiting for you to shine!

Now, let's dig a bit deeper. Imagine you're at a community college class, and you're learning about the potential of nanotechnology in clean water solutions. The instructor shares a story about a village in a developing country that lacks access to clean drinking water. The class discusses how nanotechnology could create affordable filtration systems that can purify water using tiny materials. You feel a spark inside you. "What if I could be part of a team that brings this technology to life?" you think. It's that moment of realization that can ignite a fire in your belly.

Speaking of inspiration, have you ever heard of the nanotech advancements in medicine?

Think about it—nanoparticles are being used to deliver drugs right to cancer cells, minimizing damage to healthy tissue. It's like sending a tiny, precision-guided missile straight to the target. Imagine being part of a team that develops a new treatment that saves lives. How cool would that be? You'd be at the forefront of something monumental, making a real difference in people's lives.

And let's not forget about the environment. Picture this: you're brainstorming ways to use nanotechnology to tackle climate change. You could work on creating materials that capture carbon emissions more effectively. Or maybe you're designing solar panels that are not only more efficient but also cheaper to produce. The potential is limitless, and you could be the one to push those boundaries.

But, let's keep it real. Engaging with nanotechnology isn't all rainbows and butterflies. There are challenges to face. The ethical dilemmas can be overwhelming. You might find yourself wrestling with questions like, "How do we ensure that this technology is accessible to everyone?" or "What are the long-term effects of nanomaterials on our health and environment?" These are big questions, and it's okay to feel a bit daunted. The key is to approach these challenges with an open mind and a willingness to learn.

So, how do you navigate these waters? Start by getting involved in discussions. Join forums, attend conferences, or even just chat with your peers. Share your thoughts and listen to others. You might find that the more you engage, the clearer your perspective becomes. And remember, it's perfectly fine to ask questions. Curiosity is your best friend here.

Now, let's talk about mentorship. Finding a mentor in the field can be a game-changer.

Look for someone who's been around the block a few times. They can offer guidance, share their experiences, and help you avoid common pitfalls. Plus, having someone in your corner can boost your confidence. You don't have to navigate this journey alone; there are plenty of folks who want to help.

And hey, don't forget to celebrate your wins, no matter how small. Did you complete an online course? Awesome! Share it with your network. Attend a workshop? Give yourself a pat on the back. Every step you take is a step closer to making your mark in the nanotech world.

Also, remember to keep your passion alive. It's easy to get bogged down by the nitty-gritty details or the challenges that come your way. But keep that fire burning! Find ways to stay inspired. Read books, listen to podcasts, or watch documentaries about nanotechnology.

Surround yourself with positivity and keep feeding your curiosity.

Now, let's touch on the importance of collaboration. Nanotechnology is a team sport. You'll often find yourself working with people from various backgrounds—engineers, scientists, marketers, and more. Embrace that diversity! Each person brings a unique perspective to the table, and that's where the magic happens.

Think about it: you're sitting in a brainstorming session with a group of people from different disciplines. Someone suggests a wild idea that seems totally out there. Instead of shooting it down, you all explore it further. That idea could lead to a breakthrough you never saw coming. So, keep an open mind and be ready to collaborate.

And don't forget to give back. As you grow in your journey, think about how you can help others. Maybe you can mentor someone just starting out, or volunteer at a local school to spark interest in nanotechnology among students. It's all about building a community and lifting each other up.

Now, let's circle back to responsibility. As you step into this field, remember that you're not just a cog in the machine. You're part of something bigger. Advocate for ethical practices, stay informed, and engage in discussions about the societal impacts of nanotechnology. Your voice matters, and you can help shape the future.

In the end, the nanotechnology revolution is a wild ride, and you've got a ticket to the front row. So, embrace the journey! Educate yourself, connect with others, and innovate with purpose.

The future is bright, and it's waiting for you to shine!

So, what are you waiting for? Dive in, get your hands dirty, and be a part of this incredible revolution. You've got this!

Index

Printed in Dunstable, United Kingdom